101 Problems in Calculating Derivatives Using the Chain Rule with Solutions

by

Richard Shedenhelm

INTRODUCTION

The purpose of these practice problems is to make the calculus student a master of the chain rule. The first set of 101 problems include a wide variety of function types. Each of the function types are grouped together with multiple examples to aid comprehension and confidence in the most efficient way possible. This set also minimizes notational variations (e.g., all the functions are named "$f(x)$") and unnecessary constant numbers. The next set is in random order and does *not* include any problems dealing with exponential, logarithmic, or trigonometric functions. The third set does include exponential and logarithmic functions, but not trigonometric. The fourth and final set includes all the types of functions, in random order.

Athens, Georgia, May 14, 2018.

Richard Shedenhelm

SET 1 PROBLEMS

Find the derivative $f'(x)$ of each of the following functions.

1. $f(x) = (x^2 + 3)^4$

2. $f(x) = (x^2 + 2)^{\frac{3}{2}}$

3. $f(x) = (x^3 + x^2 + 2)^5$

4. $f(x) = (x^3 + x^2 + 4)^{\frac{5}{3}}$

5. $f(x) = (x^2 + 2)^{-3}$

6. $f(x) = (x^3 + x^2 + 1)^{-\frac{1}{2}}$

7. $f(x) = \frac{1}{(x^5 + x^2)^3}$

8. $f(x) = \frac{1}{(x^4 + x)^{\frac{5}{6}}}$

9. $f(x) = \frac{1}{(x^5 + x^2)^{\frac{3}{7}}}$

10. $f(x) = \sqrt{x^2 + 3}$

11. $f(x) = \sqrt[4]{x^3 + x^2 + 4}$

12. $f(x) = \sqrt[3]{(x^2 + x)^2}$

13. $f(x) = \left(\sqrt{x^3 + 4}\right)^3$

14. $f(x) = \frac{1}{\sqrt[3]{x-1}}$

15. $f(x) = \frac{1}{\sqrt{(x^3 + x)^5}}$

16. $f(x) = \left(\frac{x^2 + 3}{x + 1}\right)^3$

17. $f(x) = \left(\frac{x - 3}{x + 8}\right)^4$

18. $f(x) = \left(\frac{x^2 - 4}{x^3 + 7}\right)^5$

19. $f(x) = \left(\frac{x^4 - x^5}{x^2 + x^3}\right)^3$

20. $f(x) = \left(x^{\frac{2}{3}} + x^{\frac{1}{2}}\right)^3$

21. $f(x) = \left(\sqrt{x} + \sqrt[3]{x^2}\right)^{\frac{4}{3}}$

22. $f(x) = \left(\sqrt[3]{x^5} + \sqrt[5]{x^{\frac{9}{2}}}\right)^{\frac{4}{3}}$

23. $f(x) = \sqrt[5]{1 + x^{\frac{2}{3}}}$

24. $f(x) = \sqrt[4]{\sqrt{x} + x^{\frac{4}{3}}}$

25. $f(x) = \left(x^{\frac{3}{4}} + x^{\frac{1}{2}}\right)^{-4}$

26. $f(x) = \frac{1}{\left(\sqrt[3]{x} + x^2\right)^{-2}}$

27. $f(x) = \frac{1}{\sqrt[4]{\sqrt{x} + x}}$

28. $f(x) = \frac{1}{\sqrt[5]{\left(\sqrt{x} + x^{\frac{1}{3}}\right)^2}}$

29. $f(x) = e^{3x}$

30. $f(x) = e^{\frac{2}{5}x}$

31. $f(x) = e^{x^2}$

32. $f(x) = e^{x^3}$

33. $f(x) = e^{e^x}$

34. $f(x) = e^{e^{2x}}$

35. $f(x) = e^{\ln(x)}$

36. $f(x) = e^{\ln(x^2)}$

37. $f(x) = e^{\arctan(x)}$

38. $f(x) = e^{\arccos(2x)}$

39. $f(x) = (e^x + e^{-x})^3$

40. $f(x) = \frac{1}{\sqrt{e^{2x} - e^{3x}}}$

41. $f(x) = \ln(2x)$

42. $f(x) = \ln(3x)$

43. $f(x) = \ln(x^2)$

44. $f(x) = \ln(3x^2)$

45. $f(x) = \ln^2(x)$

46. $f(x) = \ln^3(5x)$

47. $f(x) = \ln^4(3x^5)$

48. $f(x) = \ln(e^x)$

49. $f(x) = \ln\left(e^{x^2}\right)$

50. $f(x) = \ln(\ln(x))$

51. $f(x) = \ln(\ln(2x))$

52. $f(x) = \ln(\arctan(x))$

53. $f(x) = \ln(\arcsin(x))$

54. $f(x) = (\ln(2x) + \ln(x))^3$

55. $f(x) = \left(\ln^2(x^5) - \ln(x)\right)^{-2}$

56. $f(x) = \sin(3x)$

57. $f(x) = \cos\left(\frac{1}{2}x\right)$

58. $f(x) = \tan(3x)$

59. $f(x) = \sin(x^2)$

60. $f(x) = \cos\left(x^{\frac{3}{4}}\right)$

61. $f(x) = \tan(x^5)$

62. $f(x) = \sin^2(x)$

63. $f(x) = \cos^3(x)$

64. $f(x) = \tan^4(x)$

65. $f(x) = \sin^5(x^2)$

66. $f(x) = \cos^5(x^2)$

67. $f(x) = \tan^4(x^5)$

68. $f(x) = \sin(e^x)$

69. $f(x) = \tan(e^{3x})$

70. $f(x) = \arctan(3x)$

71. $f(x) = \arctan^2(x)$

72. $f(x) = \arctan^2(3x)$

73. $f(x) = \arcsin^3(x)$

74. $f(x) = \arcsin^3(x^2)$

75. $f(x) = \arccos(e^x)$

76. $f(x) = \sin(\arctan(x))$

77. $f(x) = \tan(\arcsin(x))$

78. $f(x) = \arctan(\sin(x))$

79. $f(x) = \arcsin(\tan(x))$

80. $f(x) = \arcsin(\arccos(x))$

81. $f(x) = \arctan(\arcsin(x))$

82. $f(x) = \ln(\sin(x))$

83. $f(x) = \ln\left(1 + \sqrt{x}\right)$

84. $f(x) = \tan(\sin(x))$

85. $f(x) = (x^3 + 1)^3(5 + x^2)^4 x^{3)^{\frac{3}{2}}}$

86. $f(x) = (x^2 + 3)^4(x^2 + 2)^{\frac{3}{2}}$

87. $f(x) = (x^2 + x)^2(-x^2 +$

88. $f(x) = \sqrt{x^3 + 1}(x^2 + 1)^4$

89. $f(x) = \sin(2x)\cos(3x)$

90. $f(x) = e^{2x}\tan^3(x)$

91. $f(x) = \dfrac{(x^2+3)^4}{(x^2+2)^{\frac{3}{2}}}$

92. $f(x) = \dfrac{e^{2x}}{\sin(3x)}$

93. $f(x) = \dfrac{\ln(4x)}{\arctan(3x)}$

94. $f(x) = e^{\cos(4x)}$

95. $f(x) = e^{\tan(\ln(3x))}$

96. $f(x) = \sin^3(\cos(2x))$

97. $f(x) = \tan^4\left(\ln\left(e^{\sin(3x)}\right)\right)$

98. $f(x) = \ln^4\left(\cos\left(e^{\sin(x^2)}\right)\right)$

99. $f(x) = \arctan^4(\cos(\ln(5x)))$

100. $f(x) = \sin\left(\sin^2\left(\sin^3(x^4)\right)\right)$

101. $f(x) = \arctan\left(\sin\left(\ln\left(e^{\sqrt{x}}\right)\right)\right)$

1. $f'(x) = 8x(x^2 + 3)^3$

2. $f'(x) = 3x\sqrt{x^2 + 2}$

3. $f'(x) = 5x(x^3 + x^2 + 2)^4(3x + 2)$

4. $f'(x) = \frac{5x}{3}(3x + 2)\sqrt[3]{(x^3 + x^2 + 4)^2}$

5. $f'(x) = \frac{-6x}{(x^2+2)^4}$

6. $f'(x) = \frac{-x(3x+2)}{2\sqrt{(x^3+x^2+1)^3}}$

7. $f'(x) = \frac{-3x(5x^3+2)}{(x^5+x^2)^4}$

8. $f'(x) = \frac{-5(4x^3+1)}{6\sqrt[6]{(x^4+x)^{11}}}$

9. $f'(x) = \frac{-3x(5x^3+2)}{7\sqrt[7]{(x^5+x^2)^{10}}}$

10. $f'(x) = \frac{x}{\sqrt{x^2+3}}$

11. $f'(x) = \frac{x(3x+2)}{4\sqrt[4]{(x^3+x^2+4)^3}}$

12. $f'(x) = \frac{2(2x+1)}{3\sqrt[3]{x^2+x}}$

13. $f'(x) = \frac{9}{2}x^2\sqrt{x^3 + 4}$

14. $f'(x) = \frac{-1}{3\sqrt[3]{(x-1)^4}}$

15. $f'(x) = \frac{-5(3x^2+1)}{2\sqrt{(x^3+x)^7}}$

16. $f'(x) = 3\left(\frac{x^2+3}{x+1}\right)^2\left(\frac{(x+3)(x-1)}{(x+1)^2}\right)$

17. $f'(x) = 44\left(\frac{x-3}{x+8}\right)^3$

18. $f'(x) = 5\left(\frac{x^2-4}{x^3+7}\right)^4\left(\frac{-x^4+14x+12x^2}{(x^3+7)^2}\right)$

19. $f'(x) = 3\left(\frac{x^4-x^5}{x^2+x^3}\right)^2\left(\frac{x^5(4-5x)(1+x)-x^5(1-x)(2+3x)}{(x^2+x^3)^2}\right)$

20. $f'(x) = 3\left(x^{\frac{2}{3}} + x^{\frac{1}{2}}\right)^2\left(\frac{2}{3}x^{-\frac{1}{3}} + \frac{1}{2}x^{-\frac{1}{2}}\right)$

21. $f'(x) = \frac{4}{3}\left(x^{\frac{1}{2}} + x^{\frac{2}{3}}\right)^{\frac{1}{3}}\left(\frac{1}{2}x^{-\frac{1}{2}} + \frac{2}{3}x^{-\frac{1}{3}}\right)$

22. $f'(x) = \frac{4}{3}\left(x^{\frac{5}{3}} + x^{\frac{9}{10}}\right)^{\frac{1}{3}}\left(\frac{5}{3}x^{\frac{2}{3}} + \frac{9}{10}x^{-\frac{1}{10}}\right)$

23. $f'(x) = \frac{2}{15}x^{-\frac{1}{3}}\left(1 + x^{\frac{2}{3}}\right)^{-\frac{4}{5}}$

24. $f'(x) = \frac{1}{4}\left(x^{\frac{1}{2}} + x^{\frac{4}{3}}\right)^{-\frac{3}{4}}\left(\frac{1}{2}x^{-\frac{1}{2}} + \frac{4}{3}x^{\frac{1}{3}}\right)$

25. $f'(x) = -4\left(x^{\frac{3}{4}} + x^{\frac{1}{2}}\right)^{-5}\left(\frac{3}{4}x^{-\frac{1}{4}} + \frac{1}{2}x^{-\frac{1}{2}}\right)$

26. $f'(x) = 2\left(x^{\frac{1}{3}} + x^2\right)\left(\frac{1}{3}x^{-\frac{2}{3}} + 2x\right)$

27. $f'(x) = -\frac{1}{4}\left(x^{\frac{1}{2}} + x\right)^{-\frac{5}{4}}\left(\frac{1}{2}x^{-\frac{1}{2}} + 1\right)$

28. $f'(x) = -\frac{2}{5}\left(x^{\frac{1}{2}} + x^{\frac{1}{3}}\right)^{-\frac{7}{5}}\left(\frac{1}{2}x^{-\frac{1}{2}} + \frac{1}{3}x^{-\frac{2}{3}}\right)$

29. $f'(x) = 3e^{3x}$

30. $f'(x) = \frac{2}{5}e^{\frac{2}{5}x}$

31. $f'(x) = 2xe^{x^2}$

32. $f'(x) = 3x^2e^{x^3}$

7

33. $f'(x) = e^{x+e^x}$

34. $f'(x) = 2e^{2x+e^{2x}}$

35. $f'(x) = 1$

36. $f'(x) = 2x$

37. $f'(x) = \frac{e^{\arctan(x)}}{1+x^2}$

38. $f'(x) = \frac{2e^{\arcsin(2x)}}{\sqrt{1-4x^2}}$

39. $f'(x) = 3(e^x + e^{-x})^2(e^x - e^{-x})$

40. $f'(x) = -\frac{1}{2}e^{2x}(2 - 3e^x)(e^{2x} - e^{3x})^{-\frac{3}{2}}$

41. $f'(x) = \frac{1}{x}$

42. $f'(x) = \frac{1}{x}$

43. $f'(x) = \frac{2}{x}$

44. $f'(x) = \frac{2}{x}$

45. $f'(x) = \frac{2\ln(x)}{x}$

46. $f'(x) = \frac{3\ln^2(5x)}{x}$

47. $f'(x) = \frac{20\ln^3(3x^5)}{x}$

48. $f'(x) = 1$

49. $f'(x) = 2x$

50. $f'(x) = \frac{1}{x\ln(x)}$

51. $f'(x) = \frac{1}{x\ln(x)}$

52. $f'(x) = \frac{1}{(1+x^2)\arctan(x)}$

53. $f'(x) = \frac{1}{\sqrt{1-x^2}\arcsin(x)}$

54. $f'(x) = \frac{6(\ln(2x)+\ln(x))^2}{x}$

55. $f'(x) = -2\big(\ln^2(x^5) - \ln(x)\big)^{-3}\left(\frac{10\ln(x^5)+1}{x}\right)$

56. $f'(x) = 3\cos(3x)$

57. $f'(x) = -\frac{1}{2}\sin\left(\frac{1}{2}x\right)$

58. $f'(x) = 3\sec^2(3x)$

59. $f'(x) = 2x\cos(x^2)$

60. $f'(x) = -\frac{3}{4}x^{\frac{-1}{4}}\sin\left(x^{\frac{3}{4}}\right)$

61. $f'(x) = 5x^4\sec^2(x^5)$

62. $f'(x) = 2\sin(x)\cos(x)$

63. $f'(x) = -3\cos^2(x)\sin(x)$

64. $f'(x) = 4\tan^3(x)\sec^2(x)$

65. $f'(x) = 10x\sin^4(x^2)\cos(x^2)$

66. $f'(x) = -10x\cos^4(x^2)\sin(x^2)$

67. $f'(x) = 20x^4 \tan^3(x^5) \sec^2(x^5)$

68. $f'(x) = e^x \cos(e^x)$

69. $f'(x) = e^{3x} \sec^2(e^{3x})$

70. $f'(x) = \frac{3}{1+9x^2}$

71. $f'(x) = \frac{2\arctan(x)}{1+x^2}$

72. $f'(x) = \frac{6\arctan(x)}{1+9x^2}$

73. $f'(x) = \frac{3\arcsin^2(x)}{\sqrt{1-x^2}}$

74. $f'(x) = \frac{6x\arcsin^2(x^2)}{\sqrt{1-x^4}}$

75. $f'(x) = \frac{-e^x}{\sqrt{1-e^{2x}}}$

76. $f'(x) = \frac{\cos(\arctan(x))}{1+x^2}$

77. $f'(x) = \frac{\sec^2(\arcsin(x))}{\sqrt{1-x^2}}$

78. $f'(x) = \frac{\cos(x)}{1+\sin^2(x)}$

79. $f'(x) = \frac{\sec^2(x)}{\sqrt{1-\tan^2(x)}}$

80. $f'(x) = \frac{1}{\sqrt{1-\arccos^2(x)}}\left(\frac{-1}{\sqrt{1-x^2}}\right)$

81. $f'(x) = \frac{1}{1+\arcsin^2(x)}\left(\frac{1}{\sqrt{1-x^2}}\right)$

82. $f'(x) = \cot(x)$

83. $f'(x) = \frac{1}{2\sqrt{x}(1+\sqrt{x})}$

84. $f'(x) = \sec^2(\sin(x))\cos(x)$

85. $f'(x) = 9x^2(x^3+1)^2(5+x^2)^4 + 8x(x^3+1)^3(5+x^2)^3$

86. $f'(x) = 8x(x^2+3)^3(x^2+2)^{\frac{3}{2}} + 3x(x^2+3)^4(x^2+2)^{\frac{1}{2}}$

87. $f'(x) = 2(x^2+x)^1(2x+1)(-x^2+x^3)^{\frac{3}{2}} +$
$+(x^2+x)^2\frac{3}{2}(-x^2+x^3)^{\frac{1}{2}}(-2x+3x)^2$

88. $f'(x) = \frac{3}{2}x^2(x^3+1)^{-\frac{1}{2}}(x^2+1)^4 + 8x(x^3+1)^{\frac{1}{2}}(x^2+1)^3$

89. $f'(x) = 2\cos(2x)\cos(3x) - 3\sin(2x)\sin(3x).$

90. $f'(x) = e^{2x}\tan^2(x)\left[2\tan(x) + 3\sec^2(x)\right]$

91. $f'(x) = \frac{8x(x^2+3)^3(x^2+2)^{\frac{3}{2}}+3x(x^2+2)^{\frac{1}{2}}(x^2+3)^4}{(x^2+2)^3}$

92. $f'(x) = \frac{2e^{2x}\sin(3x)+3\cos(3x)e^{2x}}{\sin^2(3x)}$

93. $f'(x) = \frac{\frac{\arctan(3x)}{x} - \frac{3\ln(4x)}{1+9x^2}}{\arctan^2(3x)}$

94. $f'(x) = -4\sin(4x)e^{\cos(4x)}$

95. $f'(x) = \dfrac{e^{\tan(\ln(3x))} \sec^2(\ln(3x))}{x}$

96. $f'(x) = -6\sin(2x)\cos(\cos(2x))\sin^2(\cos(2x))$

97. $f'(x) = 12\cos(3x)\tan^3\left(\ln\left(e^{\sin(3x)}\right)\right)\sec^2\left(\ln\left(e^{\sin(3x)}\right)\right)$

98. $f'(x) = -8xe^{\sin(x^2)}\cos(x^2)\tan\left(e^{\sin(x^2)}\right)\ln^3\left(\cos\left(e^{\sin(x^2)}\right)\right)$

99. $f'(x) = \dfrac{-20\sin(\ln(5x))\arctan^3(\cos(\ln(5x)))}{x(1+\cos^2(\ln(5x)))}$

100. $f'(x) =$
$24x^3\cos\left(\sin^2\left(\sin^3(x^4)\right)\right)\left(\sin\left(\sin^3(x^4)\right)\right)\left(\cos\left(\sin^3(x^4)\right)\right)\left(\sin^2(x^4)\right)(\cos(x^4))$

101. $f'(x) = \dfrac{\cos\left(\ln\left(e^{\sqrt{x}}\right)\right)}{2\sqrt{x}\left(1+\sin^2\left(\ln\left(e^{\sqrt{x}}\right)\right)\right)}$

1. $f'(x) = 4(x^2 + 3)^3 2x = 8x(x^2 + 3)^3$.

2. $f'(x) = \frac{3}{2}(x^2 + 2)^{\frac{1}{2}} 2x = 3x(x^2 + 2)^{\frac{1}{2}} = 3x\sqrt{x^2 + 2}$.

3. $f'(x) = 5(x^3 + x^2 + 2)^4(3x^2 + 2x) = 5(x^3 + x^2 + 2)^4 x(3x + 2) =$
$= 5x(x^3 + x^2 + 2)^4(3x + 2)$.

4. $f'(x) = \frac{5}{3}(x^3 + x^2 + 4)^{\frac{2}{3}}(3x^2 + 2x) = \frac{5}{3}(x^3 + x^2 + 4)^{\frac{2}{3}} x(3x + 2) =$
$= \frac{5x}{3}(3x + 2)(x^3 + x^2 + 4)^{\frac{2}{3}} = \frac{5x}{3}(3x + 2)\sqrt[3]{(x^3 + x^2 + 4)^2}$.

5. $f'(x) = -3(x^2 + 2)^{-4} 2x = -6x(x^2 + 2)^{-4} = \frac{-6x}{(x^2+2)^4}$.

6. $f'(x) = -\frac{1}{2}(x^3 + x^2 + 1)^{-\frac{3}{2}}(3x^2 + 2x) = -\frac{1}{2}(x^3 + x^2 + 1)^{-\frac{3}{2}} x(3x + 2) =$
$= \frac{-x(3x+2)}{2(x^3+x^2+1)^{\frac{3}{2}}} = \frac{-x(3x+2)}{2\sqrt{(x^3+x^2+1)^3}}$.

7. $f(x) = \frac{1}{(x^5+x^2)^3} = (x^5 + x^2)^{-3} \Rightarrow f'(x) = -3(x^5 + x^2)^{-4}(5x^4 + 2x) =$
$= -3(x^5 + x^2)^{-4} x(5x^3 + 2) = \frac{-3x(5x^3+2)}{(x^5+x^2)^4}$.

Remark: I highly recommend that if the quotient rule can be easily avoided and turned into a power rule, as in this problem, you convert the original function.

8. $f(x) = \frac{1}{(x^4+x)^{\frac{5}{6}}} = (x^4 + x)^{-\frac{5}{6}} \Rightarrow f'(x) = -\frac{5}{6}(x^4 + x)^{-\frac{11}{6}}(4x^3 + 1) =$
$= \frac{-5(4x^3+1)}{6(x^4+x)^{\frac{11}{6}}} = \frac{-5(4x^3+1)}{6\sqrt[6]{(x^4+x)^{11}}}$.

9. $f(x) = \frac{1}{(x^5+x^2)^{\frac{3}{7}}} = (x^5 + x^2)^{-\frac{3}{7}} \Rightarrow f'(x) = -\frac{3}{7}(x^5 + x^2)^{-\frac{10}{7}}(5x^4 + 2x) =$
$= -\frac{3}{7}(x^5 + x^2)^{-\frac{10}{7}} x(5x^3 + 2) = -\frac{3x}{7}(x^5 + x^2)^{-\frac{10}{7}}(5x^3 + 2) = \frac{-3x(5x^3 + 2)}{7(x^5 + x^2)^{\frac{10}{7}}} =$
$= \frac{-3x(5x^3+2)}{7\sqrt[7]{(x^5+x^2)^{10}}}$.

10. $f(x) = \sqrt{x^2 + 3} = (x^2 + 3)^{\frac{1}{2}} \Rightarrow f'(x) = \frac{1}{2}(x^2 + 3)^{-\frac{1}{2}}2x = \frac{x}{(x^2+3)^{\frac{1}{2}}} = \frac{x}{\sqrt{x^2+3}}.$

Remark: If the original function uses a radical sign, always convert it to using the exponential notation before finding the derivative function.

11. $f(x) = \sqrt[4]{x^3 + x^2 + 4} = (x^3 + x^2 + 4)^{\frac{1}{4}} \Rightarrow f'(x) = \frac{1}{4}(x^3 + x^2 + 4)^{-\frac{3}{4}}(3x^2 + 2x) =$
$= \frac{1}{4}(x^3 + x^2 + 4)^{-\frac{3}{4}}x(3x + 2) = \frac{1}{4}x(x^3 + x^2 + 4)^{-\frac{3}{4}}(3x + 2) = \frac{x(3x + 2)}{4(x^3 + x^2 + 4)^{\frac{3}{4}}} =$
$= \frac{x(3x+2)}{4\sqrt[4]{(x^3+x^2+4)^3}}.$

12. $f(x) = \sqrt[3]{(x^2 + x)^2} = ((x^2 + x)^2)^{\frac{1}{3}} = (x^2 + x)^{\frac{2}{3}} \Rightarrow$
$\Rightarrow f'(x) = \frac{2}{3}(x^2 + x)^{-\frac{1}{3}}(2x + 1) = \frac{2(2x+1)}{3(x^2+x)^{\frac{1}{3}}} = \frac{2(2x+1)}{3\sqrt[3]{x^2+x}}.$

13. $f(x) = \left(\sqrt{x^3 + 4}\right)^3 = \left((x^3 + 4)^{\frac{1}{2}}\right)^3 = (x^3 + 4)^{\frac{3}{2}} \Rightarrow f'(x) = \frac{3}{2}(x^3 + 4)^{\frac{1}{2}}3x^2 =$
$= \frac{9}{2}x^2\sqrt{x^3 + 4}.$

14. $f(x) = \frac{1}{\sqrt[3]{x-1}} = \frac{1}{(x-1)^{\frac{1}{3}}} = (x - 1)^{-\frac{1}{3}} \Rightarrow f'(x) = -\frac{1}{3}(x - 1)^{-\frac{4}{3}} = \frac{-1}{3(x-1)^{\frac{4}{3}}} = \frac{-1}{3\sqrt[3]{(x-1)^4}}.$

15. $f(x) = \frac{1}{\sqrt{(x^3+x)^5}} = \frac{1}{(x^3+x)^{\frac{5}{2}}} = (x^3 + x)^{-\frac{5}{2}} \Rightarrow f'(x) = -\frac{5}{2}(x^3 + x)^{-\frac{7}{2}}(3x^2 + 1) =$
$= \frac{-5(3x^2+1)}{2(x^3+x)^{\frac{7}{2}}} = \frac{-5(3x^2+1)}{2\sqrt{(x^3+x)^7}}.$

16. $f'(x) = 3\left(\frac{x^2+3}{x+1}\right)^2 \left(\frac{2x(x+1)-(x^2+3)}{(x+1)^2}\right) = 3\left(\frac{x^2+3}{x+1}\right)^2 \left(\frac{2x^2+2x-x^2-3}{(x+1)^2}\right) = 3\left(\frac{x^2+3}{x+1}\right)^2 \left(\frac{x^2+2x-3}{(x+1)^2}\right) =$
$= 3\left(\frac{x^2+3}{x+1}\right)^2 \left(\frac{(x+3)(x-1)}{(x+1)^2}\right).$

17. $f'(x) = 4\left(\frac{x-3}{x+8}\right)^3 \left(\frac{(x+8)-(x-3)}{(x+8)^2}\right) = 4\left(\frac{x-3}{x+8}\right)^3 \left(\frac{x+8-x+3}{(x+8)^2}\right) = 4\left(\frac{x-3}{x+8}\right)^3 \left(\frac{11}{(x+8)^2}\right) = 44\left(\frac{x-3}{x+8}\right)^3.$

18. $f'(x) = 5\left(\frac{x^2-4}{x^3+7}\right)^4 \left(\frac{2x(x^3+7)-(x^2-4)3x^2}{(x^3+7)^2}\right) = 5\left(\frac{x^2-4}{x^3+7}\right)^4 \left(\frac{2x^4+14x-3x^4+12x^2}{(x^3+7)^2}\right) =$
$= 5\left(\frac{x^2-4}{x^3+7}\right)^4 \left(\frac{-x^4+14x+12x^2}{(x^3+7)^2}\right).$

19. $f'(x) = 3\left(\frac{x^4-x^5}{x^2+x^3}\right)^2 \left(\frac{(4x^3-5x^4)(x^2+x^3)-(x^4-x^5)(2x+3x^2)}{(x^2+x^3)^2}\right) =$

$= 3\left(\frac{x^4-x^5}{x^2+x^3}\right)^2 \left(\frac{x^3(4-5x)x^2(1+x)-x^4(1-x)x(2+3x)}{(x^2+x^3)^2}\right) =$

$= 3\left(\frac{x^4-x^5}{x^2+x^3}\right)^2 \left(\frac{x^5(4-5x)(1+x)-x^5(1-x)(2+3x)}{(x^2+x^3)^2}\right).$

20. $f'(x) = 3\left(x^{\frac{2}{3}} + x^{\frac{1}{2}}\right)^2 \left(\frac{2}{3}x^{-\frac{1}{3}} + \frac{1}{2}x^{-\frac{1}{2}}\right).$

21. $f(x) = \left(\sqrt{x} + \sqrt[3]{x^2}\right)^{\frac{4}{3}} = \left(x^{\frac{1}{2}} + x^{\frac{2}{3}}\right)^{\frac{4}{3}} \Rightarrow f'(x) = \frac{4}{3}\left(x^{\frac{1}{2}} + x^{\frac{2}{3}}\right)^{\frac{1}{3}} \left(\frac{1}{2}x^{-\frac{1}{2}} + \frac{2}{3}x^{-\frac{1}{3}}\right).$

22. $f(x) = \left(\sqrt[3]{x^5} + \sqrt[5]{x^{\frac{9}{2}}}\right)^{\frac{4}{3}} = \left((x^5)^{\frac{1}{3}} + \left(x^{\frac{9}{2}}\right)^{\frac{1}{5}}\right)^{\frac{4}{3}} = \left(x^{\frac{5}{3}} + x^{\frac{9}{10}}\right)^{\frac{4}{3}} \Rightarrow$

$\Rightarrow f'(x) = \frac{4}{3}\left(x^{\frac{5}{3}} + x^{\frac{9}{10}}\right)^{\frac{1}{3}} \left(\frac{5}{3}x^{\frac{2}{3}} + \frac{9}{10}x^{-\frac{1}{10}}\right).$

23. $f(x) = \sqrt[5]{1 + x^{\frac{2}{3}}} = \left(1 + x^{\frac{2}{3}}\right)^{\frac{1}{5}} \Rightarrow f'(x) = \frac{1}{5}\left(1 + x^{\frac{2}{3}}\right)^{-\frac{4}{5}}\frac{2}{3}x^{-\frac{1}{3}} = \frac{2}{15}x^{-\frac{1}{3}}\left(1 + x^{\frac{2}{3}}\right)^{-\frac{4}{5}}.$

24. $f(x) = \sqrt[4]{\sqrt{x} + x^{\frac{4}{3}}} = \left(x^{\frac{1}{2}} + x^{\frac{4}{3}}\right)^{\frac{1}{4}} \Rightarrow f'(x) = \frac{1}{4}\left(x^{\frac{1}{2}} + x^{\frac{4}{3}}\right)^{-\frac{3}{4}} \left(\frac{1}{2}x^{-\frac{1}{2}} + \frac{4}{3}x^{\frac{1}{3}}\right).$

25. $f'(x) = -4\left(x^{\frac{3}{4}} + x^{\frac{1}{2}}\right)^{-5} \left(\frac{3}{4}x^{-\frac{1}{4}} + \frac{1}{2}x^{-\frac{1}{2}}\right).$

26. $f(x) = \frac{1}{\left(\sqrt[3]{x} + x^2\right)^{-2}} = \left(x^{\frac{1}{3}} + x^2\right)^2 \Rightarrow f'(x) = 2\left(x^{\frac{1}{3}} + x^2\right)\left(\frac{1}{3}x^{-\frac{2}{3}} + 2x\right).$

27. $f(x) = \frac{1}{\sqrt[4]{\sqrt{x}+x}} = \frac{1}{\left(x^{\frac{1}{2}}+x\right)^{\frac{1}{4}}} = \left(x^{\frac{1}{2}} + x\right)^{-\frac{1}{4}} \Rightarrow f'(x) = -\frac{1}{4}\left(x^{\frac{1}{2}} + x\right)^{-\frac{5}{4}} \left(\frac{1}{2}x^{-\frac{1}{2}} + 1\right).$

28. $f(x) = \frac{1}{\sqrt[5]{\left(\sqrt{x}+x^{\frac{1}{3}}\right)^2}} = \frac{1}{\left(\left(x^{\frac{1}{2}}+x^{\frac{1}{3}}\right)^2\right)^{\frac{1}{5}}} = \frac{1}{\left(x^{\frac{1}{2}}+x^{\frac{1}{3}}\right)^{\frac{2}{5}}} = \left(x^{\frac{1}{2}} + x^{\frac{1}{3}}\right)^{-\frac{2}{5}} \Rightarrow$

$\Rightarrow f'(x) = -\frac{2}{5}\left(x^{\frac{1}{2}} + x^{\frac{1}{3}}\right)^{-\frac{7}{5}} \left(\frac{1}{2}x^{-\frac{1}{2}} + \frac{1}{3}x^{-\frac{2}{3}}\right).$

29. $f'(x) = e^{3x}(3) = 3e^{3x}.$

13

30. $f'(x) = e^{\frac{2}{5}x} \left(\frac{2}{5}\right) = \frac{2}{5} e^{\frac{2}{5}x}.$

31. $f'(x) = e^{x^2}(2x) = 2xe^{x^2}.$

32. $f'(x) = e^{x^3}(3x^2) = 3x^2 e^{x^3}.$

33. $f'(x) = e^{e^x}(e^x) = e^x e^{e^x} = e^{x+e^x}.$

34. $f'(x) = e^{e^{2x}}(e^{2x})(2) = 2e^{2x} e^{e^{2x}} = 2e^{2x+e^{2x}}.$

35. $f'(x) = e^{\ln(x)} \left(\frac{1}{x}\right) = \frac{e^{\ln(x)}}{x} = \frac{x}{x} = 1.$

36. $f'(x) = e^{\ln(x^2)} \left(\frac{1}{x^2}\right)(2x) = \frac{2e^{\ln(x^2)}}{x} = \frac{2x^2}{x} = 2x.$

37. $f'(x) = e^{\arctan(x)} \left(\frac{1}{1+x^2}\right) = \frac{e^{\arctan(x)}}{1+x^2}.$

38. $f'(x) = e^{\arcsin(2x)} \left(\frac{1}{\sqrt{1+(2x)^2}}\right)(2) = \frac{2e^{\arcsin(2x)}}{\sqrt{1+4x^2}}.$

39. $f'(x) = 3(e^x + e^{-x})^2 \left(e^x + e^{-x}(-1)\right) = 3(e^x + e^{-x})^2(e^x - e^{-x}).$

40. $f(x) = \frac{1}{\sqrt{e^{2x}-e^{3x}}} = \frac{1}{(e^{2x}-e^{3x})^{\frac{1}{2}}} = (e^{2x} - e^{3x})^{-\frac{1}{2}} \Rightarrow$

$f'(x) = -\frac{1}{2}(e^{2x} - e^{3x})^{-\frac{3}{2}}\left(e^{2x}(2) - e^{3x}(3)\right) = -\frac{1}{2}(e^{2x} - e^{3x})^{-\frac{3}{2}}(2e^{2x} - 3e^{3x}) =$

$= -\frac{1}{2}e^{2x}(2 - 3e^x)(e^{2x} - e^{3x})^{-\frac{3}{2}}.$

41. $f'(x) = \frac{1}{2x}(2) = \frac{2}{2x} = \frac{1}{x}.$
    Alternate Solution:
Using the logarithmic properties,
$f(x) = \ln(2x) = \ln(2) + \ln(x) \Rightarrow f'(x) = 0 + \frac{1}{x} = \frac{1}{x}.$

42. $f'(x) = \frac{1}{3x}(3) = \frac{3}{3x} = \frac{1}{x}.$
    Alternate Solution:
Using the logarithmic properties,
$f(x) = \ln(3x) = \ln(3) + \ln(x) \Rightarrow f'(x) = 0 + \frac{1}{x} = \frac{1}{x}.$

43. $f'(x) = \frac{1}{x^2}(2x) = \frac{2x}{x^2} = \frac{2}{x}$.

      Alternate Solution:

Using the logarithmic properties,

$f(x) = \ln(x^2) = 2\ln(x) \Rightarrow f'(x) = 2\left(\frac{1}{x}\right) = \frac{2}{x}$.

44. $f'(x) = \frac{1}{3x^2}(6x) = \frac{6x}{3x^2} = \frac{2}{x}$.

      Alternate Solution:

Using the logarithmic properties,

$f(x) = \ln(3x^2) = \ln(3) + \ln(x^2) = \ln(3) + 2\ln(x) \Rightarrow f'(x) = 0 + 2\left(\frac{1}{x}\right) = \frac{2}{x}$.

45. $f(x) = \ln^2(x) = [\ln(x)]^2 \Rightarrow f'(x) = 2\ln(x)\frac{1}{x} = \frac{2\ln(x)}{x}$.

Remark: Converting the original function as I have done here clarifies what the "outside" function is as opposed to the "inside" function.

46. $f(x) = \ln^3(5x) = [\ln(5x)]^3 \Rightarrow f'(x) = 3[\ln(5x)]^2\frac{1}{5x}(5) = 3\ln^2(5x)\frac{1}{5x}(5) = \frac{3\ln^2(5x)}{x}$.

47. $f(x) = \ln^4(3x^5) = [\ln(3x^5)]^4 \Rightarrow f'(x) = 4\ln^3(3x^5)\frac{1}{3x^5}(15x^4) = \frac{60x^4\ln^3(3x^5)}{3x^5} =$

$= \frac{20\ln^3(3x^5)}{x}$.

48. $f'(x) = \frac{1}{e^x}(e^x) = \frac{e^x}{e^x} = 1$.

      Alternate Solution:

Using the logarithmic properties,

$f(x) = \ln(e^x) = x\ln(e) = x(1) = x \Rightarrow f'(x) = 1$.

49. $f'(x) = \frac{1}{e^{x^2}}(e^{x^2})(2x) = \frac{2xe^{x^2}}{e^{x^2}} = 2x$.

      Alternate Solution:

Using the logarithmic properties,

$f(x) = \ln(e^{x^2}) = x^2\ln(e) = x^2(1) = x^2 \Rightarrow f'(x) = 2x$.

50. $f'(x) = \frac{1}{\ln(x)}\left(\frac{1}{x}\right) = \frac{1}{x\ln(x)}$.

51. $f'(x) = \frac{1}{\ln(2x)}\left(\frac{1}{2x}\right)(2) = \frac{1}{x\ln(x)}$.

52. $f'(x) = \frac{1}{\arctan(x)}\left(\frac{1}{1+x^2}\right) = \frac{1}{(1+x^2)\arctan(x)}$.

53. $f'(x) = \frac{1}{\arcsin(x)}\left(\frac{1}{\sqrt{1-x^2}}\right) = \frac{1}{\sqrt{1-x^2}\,\arcsin(x)}.$

54. $f'(x) = 3(\ln(2x) + \ln(x))^2 \left(\frac{2}{2x} + \frac{1}{x}\right) = 3(\ln(2x) + \ln(x))^2 \left(\frac{1}{x} + \frac{1}{x}\right) =$
$= 3(\ln(2x) + \ln(x))^2 \left(\frac{2}{x}\right) = \frac{6(\ln(2x)+\ln(x))^2}{x}.$

55. $f'(x) = -2\left(\ln{}^2(x^5) - \ln(x)\right)^{-3}\left(2\ln(x^5)\frac{5x^4}{x^5} + \frac{1}{x}\right) =$
$= -2\left(\ln{}^2(x^5) - \ln(x)\right)^{-3}\left(\frac{10}{x}\ln(x^5) + \frac{1}{x}\right) = -2\left(\ln{}^2(x^5) - \ln(x)\right)^{-3}\left(\frac{10\ln(x^5)+1}{x}\right).$

56. $f'(x) = \cos(3x)(3) = 3\cos(3x).$

57. $f'(x) = -\sin\left(\frac{1}{2}x\right)\left(\frac{1}{2}\right) = -\frac{1}{2}\sin\left(\frac{1}{2}x\right).$

58. $f'(x) = \sec{}^2(3x)(3) = 3\sec{}^2(3x).$

59. $f'(x) = \cos(x^2)(2x) = 2x\cos(x^2).$

60. $f'(x) = -\sin\left(x^{\frac{3}{4}}\right)\frac{3}{4}x^{\frac{-1}{4}} = -\frac{3}{4}x^{\frac{-1}{4}}\sin\left(x^{\frac{3}{4}}\right).$

61. $f'(x) = \sec{}^2(x^5)5x^4 = 5x^4\sec{}^2(x^5).$

62. $f(x) = \sin{}^2(x) = [\sin(x)]^2 \Rightarrow f'(x) = 2\sin(x)\cos(x).$

Remark: The conversion of the original function, as in problems 45-47, is done to distinguish the "outside" from the "inside" function.

63. $f(x) = \cos{}^3(x) = [\cos(x)]^3 \Rightarrow f'(x) = 3[\cos(x)]^2(-\sin(x)) = -3\cos{}^2(x)\sin(x).$

64. $f(x) = \tan{}^4(x) = [\tan(x)]^4 \Rightarrow f'(x) = 4\tan{}^3(x)\sec{}^2(x).$

65. $f(x) = \sin{}^5(x^2) = [\sin(x^2)]^5 \Rightarrow f'(x) = 5\sin{}^4(x^2)\cos(x^2)2x =$
$= 10x\sin{}^4(x^2)\cos(x^2).$

66. $f(x) = \cos{}^5(x^2) = [\cos(x^2)]^5 \Rightarrow f'(x) = -5\cos{}^4(x^2)\sin(x^2)2x =$
$= -10x\cos{}^4(x^2)\sin(x^2).$

67. $f(x) = \tan{}^4(x^5) = [\tan(x^5)]^4 \Rightarrow f'(x) = 4\tan{}^3(x^5)\sec{}^2(x^5)5x^4 =$
$= 20x^4\tan{}^3(x^5)\sec{}^2(x^5).$

68. $f'(x) = \cos(e^x)(e^x) = e^x\cos(e^x).$

69. $f'(x) = \sec^2(e^{3x})(e^{3x})(3) = 3e^{3x}\sec^2(e^{3x})$.

70. $f'(x) = \dfrac{3}{1+(3x)^2} = \dfrac{3}{1+9x^2}$.

71. $f(x) = \arctan^2(x) = [\arctan(x)]^2 \Rightarrow f'(x) = 2\arctan(x)\left(\dfrac{1}{1+x^2}\right) = \dfrac{2\arctan(x)}{1+x^2}$.

72. $f(x) = \arctan^2(3x) = [\arctan(3x)]^2 \Rightarrow f'(x) = 2\arctan(3x)\left(\dfrac{3}{1+(3x)^2}\right) = \dfrac{6\arctan(x)}{1+9x^2}$.

73. $f(x) = \arcsin^3(x) = [\arcsin(x)]^3 \Rightarrow f'(x) = 3\arcsin^2(x)\left(\dfrac{1}{\sqrt{1-x^2}}\right) = \dfrac{3\arcsin^2(x)}{\sqrt{1-x^2}}$.

74. $f(x) = \arcsin^3(x^2) = [\arcsin(x^2)]^3 \Rightarrow f'(x) = 3\arcsin^2(x^2)\left(\dfrac{2x}{\sqrt{1-x^4}}\right) = \dfrac{6x\arcsin^2(x^2)}{\sqrt{1-x^4}}$.

75. $f'(x) = \dfrac{-1}{\sqrt{1-(e^x)^2}}(e^x) = \dfrac{-e^x}{\sqrt{1-e^{2x}}}$.

76. $f'(x) = \cos(\arctan(x))\left(\dfrac{1}{1+x^2}\right) = \dfrac{\cos(\arctan(x))}{1+x^2}$.

77. $f'(x) = \sec^2(\arcsin(x))\left(\dfrac{1}{\sqrt{1-x^2}}\right) = \dfrac{\sec^2(\arcsin(x))}{\sqrt{1-x^2}}$.

78. $f'(x) = \dfrac{1}{1+\sin^2(x)}(\cos(x)) = \dfrac{\cos(x)}{1+\sin^2(x)}$.

79. $f'(x) = \dfrac{1}{\sqrt{1-\tan^2(x)}}\left(\sec^2(x)\right) = \dfrac{\sec^2(x)}{\sqrt{1-\tan^2(x)}}$.

80. $f'(x) = \dfrac{1}{\sqrt{1-\arccos^2(x)}}\left(\dfrac{-1}{\sqrt{1-x^2}}\right)$.

81. $f'(x) = \dfrac{1}{1+\arcsin^2(x)}\left(\dfrac{1}{\sqrt{1-x^2}}\right)$.

82. $f'(x) = \dfrac{\cos(x)}{\sin(x)} = \cot(x)$.

83. $f(x) = \ln(1+\sqrt{x}) = \ln\left(1+x^{\frac{1}{2}}\right) \Rightarrow f'(x) = \dfrac{\frac{1}{2}x^{-\frac{1}{2}}}{1+x^{\frac{1}{2}}} = \dfrac{1}{2x^{\frac{1}{2}}\left(1+x^{\frac{1}{2}}\right)} = \dfrac{1}{2\sqrt{x}(1+\sqrt{x})}$.

84. $f'(x) = \sec^2(\sin(x))\cos(x)$.

85. $f'(x) = 3(x^3+1)^2 3x^2(5+x^2)^4 + (x^3+1)^3 4(5+x^2)^3 2x = $
$= 9x^2(x^3+1)^2(5+x^2)^4 + 8x(x^3+1)^3(5+x^2)^3$.

86. $f'(x) = 4(x^2 + 3)^3 2x(x^2 + 2)^{\frac{3}{2}} + (x^2 + 3)^4 \frac{3}{2}(x^2 + 2)^{\frac{1}{2}} 2x =$
$= 8x(x^2 + 3)^3(x^2 + 2)^{\frac{3}{2}} + 3x(x^2 + 3)^4(x^2 + 2)^{\frac{1}{2}}.$

87. $f'(x) = 2(x^2 + x)^1(2x + 1)(-x^2 + x^3)^{\frac{3}{2}} +$
$+ (x^2 + x)^2 \frac{3}{2}(-x^2 + x^3)^{\frac{1}{2}}(-2x + 3x)^2.$

88. $f(x) = \sqrt{x^3 + 1}(x^2 + 1)^4 = (x^3 + 1)^{\frac{1}{2}}(x^2 + 1)^4 \Rightarrow$
$\Rightarrow f'(x) = \frac{1}{2}(x^3 + 1)^{-\frac{1}{2}}(3x^2)(x^2 + 1)^4 + (x^3 + 1)^{\frac{1}{2}} 4(x^2 + 1)^3(2x) =$
$= \frac{3}{2}x^2(x^3 + 1)^{-\frac{1}{2}}(x^2 + 1)^4 + 8x(x^3 + 1)^{\frac{1}{2}}(x^2 + 1)^3.$

89. $f'(x) = 2\cos(2x)\cos(3x) - 3\sin(2x)\sin(3x).$

90. $f'(x) = 2e^{2x}\tan^3(x) + 3e^{2x}\tan^2(x)\sec^2(x) = e^{2x}\tan^2(x)\left[2\tan(x) + 3\sec^2(x)\right].$

91. $f'(x) = \dfrac{4(x^2+3)^3(2x)(x^2+2)^{\frac{3}{2}} + \frac{3}{2}(x^2+2)^{\frac{1}{2}}(2x)(x^2+3)^4}{(x^2+2)^3} = \dfrac{8x(x^2+3)^3(x^2+2)^{\frac{3}{2}} + 3x(x^2+2)^{\frac{1}{2}}(x^2+3)^4}{(x^2+2)^3}.$

92. $f'(x) = \dfrac{2e^{2x}\sin(3x) + 3\cos(3x)e^{2x}}{\sin^2(3x)}.$

93. $f'(x) = \dfrac{\frac{1}{4x}(4)\arctan(3x) - \ln(4x)\frac{1}{1+9x^2}(3)}{\arctan^2(3x)} = \dfrac{\frac{\arctan(3x)}{x} - \frac{3\ln(4x)}{1+9x^2}}{\arctan^2(3x)}.$

94. $f'(x) = e^{\cos(4x)}(-\sin(4x))4 = -4\sin(4x)e^{\cos(4x)}.$

95. $f'(x) = e^{\tan(\ln(3x))}\sec^2(\ln(3x))\left(\frac{1}{3x}\right)(3) = \dfrac{e^{\tan(\ln(3x))}\sec^2(\ln(3x))}{x}.$

96. $f'(x) = 3\sin^2(\cos(2x))\cos(\cos(2x))(-\sin(2x))(2) =$
$= -6\sin(2x)\cos(\cos(2x))\sin^2(\cos(2x)).$

97. $f(x) = \tan^4\left(\ln\left(e^{\sin(3x)}\right)\right) = \left(\tan\left(\ln\left(e^{\sin(3x)}\right)\right)\right)^4 \Rightarrow$

$$f'(x) = 4\left(\tan\left(\ln\left(e^{\sin(3x)}\right)\right)\right)^3 \sec^2\left(\ln\left(e^{\sin(3x)}\right)\right)\left(\frac{1}{e^{\sin(3x)}}\right)\left(e^{\sin(3x)}\right)(\cos(3x))(3) =$$

$$= 4\left(\tan\left(\ln\left(e^{\sin(3x)}\right)\right)\right)^3 \sec^2\left(\ln\left(e^{\sin(3x)}\right)\right)\left(\frac{e^{\sin(3x)}\, 3\cos(3x)}{e^{\sin(3x)}}\right) =$$

$$= 4\tan^3\left(\ln\left(e^{\sin(3x)}\right)\right)\sec^2\left(\ln\left(e^{\sin(3x)}\right)\right)\left(\frac{e^{\sin(3x)}\, 3\cos(3x)}{e^{\sin(3x)}}\right) =$$

$$= 4\tan^3\left(\ln\left(e^{\sin(3x)}\right)\right)\sec^2\left(\ln\left(e^{\sin(3x)}\right)\right)3\cos(3x) = 12\cos(3x)\tan^3\left(\ln\left(e^{\sin(3x)}\right)\right)\sec^2\left(\ln\left(e^{\sin(3x)}\right)\right).$$

98. $f(x) = \ln^4\left(\cos\left(e^{\sin(x^2)}\right)\right) = \left(\ln\left(\cos\left(e^{\sin(x^2)}\right)\right)\right)^4 \Rightarrow$

$$f'(x) = 4\left(\ln\left(\cos\left(e^{\sin(x^2)}\right)\right)\right)^3 \frac{1}{\cos\left(e^{\sin(x^2)}\right)}\left(-\sin\left(e^{\sin(x^2)}\right)\right)e^{\sin(x^2)}(\cos(x^2))(2x) =$$

$$= \frac{-8xe^{\sin(x^2)}\cos(x^2)\sin\left(e^{\sin(x^2)}\right)\left(\ln\left(\cos\left(e^{\sin(x^2)}\right)\right)\right)^3}{\cos\left(e^{\sin(x^2)}\right)} =$$

$$= \frac{-8xe^{\sin(x^2)}\cos(x^2)\sin\left(e^{\sin(x^2)}\right)\ln^3\left(\cos\left(e^{\sin(x^2)}\right)\right)}{\cos\left(e^{\sin(x^2)}\right)} =$$

$$= -8xe^{\sin(x^2)}\cos(x^2)\tan\left(e^{\sin(x^2)}\right)\ln^3\left(\cos\left(e^{\sin(x^2)}\right)\right).$$

99. $f(x) = \arctan^4(\cos(\ln(5x))) = \left(\arctan(\cos(\ln(5x)))\right)^4 \Rightarrow$

$$f'(x) = 4\left(\arctan(\cos(\ln(5x)))\right)^3 \frac{1}{1+\cos^2(\ln(5x))}(-\sin(\ln(5x)))\left(\frac{1}{5x}\right)(5) =$$

$$= \frac{-20\sin(\ln(5x))\left(\arctan(\cos(\ln(5x)))\right)^3}{x(1+\cos^2(\ln(5x)))} = \frac{-20\sin(\ln(5x))\arctan^3(\cos(\ln(5x)))}{x(1+\cos^2(\ln(5x)))}.$$

100. $f(x) = \sin\left(\sin^2(\sin^3(x^4))\right) = \sin\left([\sin(\sin^3(x^4))]^2\right) = \sin([\sin([\sin(x^4)]^3)]^2) \Rightarrow$

$f'(x) =$

$= \cos\left(\sin^2(\sin^3(x^4))\right)\left(2\sin(\sin^3(x^4))\right)\left(\cos(\sin^3(x^4))\right)\left(3\sin^2(x^4)\right)(\cos(x^4))4x^3 =$

$= 24x^3 \cos\left(\sin^2(\sin^3(x^4))\right)\left(\sin(\sin^3(x^4))\right)\left(\cos(\sin^3(x^4))\right)(\sin^2(x^4))(\cos(x^4)).$

101. $f(x) = \arctan\left(\sin(\ln(e^{\sqrt{x}}))\right) = \arctan\left(\sin\left(\ln\left(e^{x^{\frac{1}{2}}}\right)\right)\right) \Rightarrow$

$f'(x) = \dfrac{1}{1 + \sin^2\left(\ln\left(e^{x^{\frac{1}{2}}}\right)\right)}\left(\cos\left(\ln\left(e^{x^{\frac{1}{2}}}\right)\right)\right)\left(\dfrac{1}{e^{x^{\frac{1}{2}}}}\right)\left(e^{x^{\frac{1}{2}}}\right)\left(\dfrac{1}{2}x^{-\frac{1}{2}}\right) =$

$= \dfrac{\cos\left(\ln\left(e^{x^{\frac{1}{2}}}\right)\right)}{2x^{\frac{1}{2}}\left(1 + \sin^2\left(\ln\left(e^{x^{\frac{1}{2}}}\right)\right)\right)} = \dfrac{\cos(\ln(e^{\sqrt{x}}))}{2\sqrt{x}\left(1 + \sin^2(\ln(e^{\sqrt{x}}))\right)}.$

SET 2 PROBLEMS

Find the derivative $f'(x)$ of each of the following functions.

1. $f(x) = \left(\frac{x^2+3}{x+1}\right)^3$

2. $f(x) = \sqrt{x^3+1}(x^2+1)^4$

3. $f(x) = \sqrt[4]{\sqrt{x} + x^{\frac{4}{3}}}$

4. $f(x) = \sqrt[5]{1 + x^{\frac{2}{3}}}$

5. $f(x) = (x^3 + x^2 + 4)^{\frac{5}{3}}$

6. $f(x) = \sqrt[3]{(x^2 + x)^2}$

7. $f(x) = \left(\frac{x^4 - x^5}{x^2 + x^3}\right)^3$

8. $f(x) = (x^2 + x)^2(-x^2 + x^3)^{\frac{3}{2}}$

9. $f(x) = \left(\frac{x-3}{x+8}\right)^4$

10. $f(x) = \left(\frac{x^2 - 4}{x^3 + 7}\right)^5$

11. $f(x) = (x^2 + 3)^4$

12. $f(x) = \frac{1}{\sqrt[3]{x-1}}$

13. $f(x) = \left(\sqrt{x^3 + 4}\right)^3$

14. $f(x) = \frac{1}{\left(\sqrt[3]{x} + x^2\right)^{-2}}$

15. $f(x) = \frac{1}{\sqrt{(x^3+x)^5}}$

16. $f(x) = \left(x^{\frac{2}{3}} + x^{\frac{1}{2}}\right)^3$

17. $f(x) = (x^3 + x^2 + 2)^5$

18. $f(x) = \sqrt{x^2 + 3}$

19. $f(x) = (x^2 + 2)^{-3}$

20. $f(x) = \frac{1}{\sqrt[4]{\sqrt{x}+x}}$

21. $f(x) = \left(\sqrt{x} + \sqrt[3]{x^2}\right)^{\frac{4}{3}}$

22. $f(x) = \sqrt[4]{x^3 + x^2 + 4}$

23. $f(x) = \frac{1}{(x^5+x^2)^3}$

24. $f(x) = \frac{1}{(x^5+x^2)^{\frac{3}{7}}}$

25. $f(x) = (x^3 + 1)^3(5 + x^2)^4$

26. $f(x) = (x^3 + x^2 + 1)^{-\frac{1}{2}}$

27. $f(x) = \left(\sqrt[3]{x^5} + \sqrt[5]{x^{\frac{9}{2}}}\right)^{\frac{4}{3}}$

28. $f(x) = \frac{1}{\sqrt[5]{\left(\sqrt{x}+x^{\frac{1}{3}}\right)^2}}$

29. $f(x) = \left(x^{\frac{3}{4}} + x^{\frac{1}{2}}\right)^{-4}$

30. $f(x) = (x^2 + 3)^4(x^2 + 2)^{\frac{3}{2}}$

31. $f(x) = \frac{1}{(x^4+x)^{\frac{5}{6}}}$

32. $f(x) = (x^2 + 2)^{\frac{3}{2}}$

1. $f'(x) = 3\left(\frac{x^2+3}{x+1}\right)^2 \left(\frac{(x+3)(x-1)}{(x+1)^2}\right)$

2. $f'(x) = \frac{3}{2}x^2(x^3+1)^{-\frac{1}{2}}(x^2+1)^4 + 8x(x^3+1)^{\frac{1}{2}}(x^2+1)^3$

3. $f'(x) = \frac{1}{4}\left(x^{\frac{1}{2}} + x^{\frac{4}{3}}\right)^{-\frac{3}{4}}\left(\frac{1}{2}x^{-\frac{1}{2}} + \frac{4}{3}x^{\frac{1}{3}}\right)$

4. $f'(x) = \frac{2}{15}x^{-\frac{1}{3}}\left(1 + x^{\frac{2}{3}}\right)^{-\frac{4}{5}}$

5. $f'(x) = \frac{5x}{3}(3x+2)\sqrt[3]{(x^3+x^2+4)^2}$

6. $f'(x) = \frac{2(2x+1)}{3\sqrt[3]{x^2+x}}$

7. $f'(x) = 3\left(\frac{x^4-x^5}{x^2+x^3}\right)^2 \left(\frac{x^5(4-5x)(1+x)-x^5(1-x)(2+3x)}{(x^2+x^3)^2}\right)$

8. $f'(x) = 2(x^2+x)^1(2x+1)(-x^2+x^3)^{\frac{3}{2}} +$
   $+ (x^2+x)^2\frac{3}{2}(-x^2+x^3)^{\frac{1}{2}}(-2x+3x)^2$

9. $f'(x) = 44\left(\frac{x-3}{x+8}\right)^3$

10. $f'(x) = 5\left(\frac{x^2-4}{x^3+7}\right)^4 \left(\frac{-x^4+14x+12x^2}{(x^3+7)^2}\right)$

11. $f'(x) = 8x(x^2+3)^3$

12. $f'(x) = \frac{-1}{3\sqrt[3]{(x-1)^4}}$

13. $f'(x) = \frac{9}{2}x^2\sqrt{x^3+4}$

14. $f'(x) = 2\left(x^{\frac{1}{3}} + x^2\right)\left(\frac{1}{3}x^{-\frac{2}{3}} + 2x\right)$

15. $f'(x) = \frac{-5(3x^2+1)}{2\sqrt{(x^3+x)^7}}$

16. $f'(x) = 3\left(x^{\frac{2}{3}} + x^{\frac{1}{2}}\right)^2 \left(\frac{2}{3}x^{-\frac{1}{3}} + \frac{1}{2}x^{-\frac{1}{2}}\right)$

17. $f'(x) = 5x(x^3+x^2+2)^4(3x+2)$

18. $f'(x) = \frac{x}{\sqrt{x^2+3}}$

19. $f'(x) = \frac{-6x}{(x^2+2)^4}$

20. $f'(x) = -\frac{1}{4}\left(x^{\frac{1}{2}} + x\right)^{-\frac{5}{4}}\left(\frac{1}{2}x^{-\frac{1}{2}} + 1\right)$

21. $f'(x) = \frac{4}{3}\left(x^{\frac{1}{2}} + x^{\frac{2}{3}}\right)^{\frac{1}{3}}\left(\frac{1}{2}x^{-\frac{1}{2}} + \frac{2}{3}x^{-\frac{1}{3}}\right)$

22. $f'(x) = \frac{x(3x+2)}{4\sqrt[4]{(x^3+x^2+4)^3}}$

23. $f'(x) = \frac{-3x(5x^3+2)}{(x^5+x^2)^4}$

24. $f'(x) = \frac{-3x(5x^3+2)}{7\sqrt[7]{(x^5+x^2)^{10}}}$

25. $f'(x) = 9x^2(x^3+1)^2(5+x^2)^4 + 8x(x^3+1)^3(5+x^2)^3$

26. $f'(x) = \frac{-x(3x+2)}{2\sqrt{(x^3+x^2+1)^3}}$

27. $f'(x) = \frac{4}{3}\left(x^{\frac{5}{3}} + x^{\frac{9}{10}}\right)^{\frac{1}{3}}\left(\frac{5}{3}x^{\frac{2}{3}} + \frac{9}{10}x^{-\frac{1}{10}}\right)$

28. $f'(x) = -\frac{2}{5}\left(x^{\frac{1}{2}} + x^{\frac{1}{3}}\right)^{-\frac{7}{5}}\left(\frac{1}{2}x^{-\frac{1}{2}} + \frac{1}{3}x^{-\frac{2}{3}}\right)$

29. $f'(x) = -4\left(x^{\frac{3}{4}} + x^{\frac{1}{2}}\right)^{-5}\left(\frac{3}{4}x^{-\frac{1}{4}} + \frac{1}{2}x^{-\frac{1}{2}}\right)$

30. $f'(x) = 8x(x^2+3)^3(x^2+2)^{\frac{3}{2}} + 3x(x^2+3)^4(x^2+2)^{\frac{1}{2}}$

31. $f'(x) = \frac{-5(4x^3+1)}{6\sqrt[6]{(x^4+x)^{11}}}$

32. $f'(x) = 3x\sqrt{x^2+2}$

SET 3 PROBLEMS

Find the derivative $f'(x)$ of each of the following functions.

1. $f(x) = \sqrt[3]{(x^2 + x)^2}$

2. $f(x) = e^{e^{2x}}$

3. $f(x) = \sqrt{x^3 + 1}(x^2 + 1)^4$

4. $f(x) = \ln(3x^2)$

5. $f(x) = \sqrt[4]{\sqrt{x} + x^{\frac{4}{3}}}$

6. $f(x) = \left(\ln^2(x^5) - \ln(x)\right)^{-2}$

7. $f(x) = e^{\frac{2}{5}x}$

8. $f(x) = \sqrt[5]{1 + x^{\frac{2}{3}}}$

9. $f(x) = (x^3 + x^2 + 4)^{\frac{5}{3}}$

10. $f(x) = (e^x + e^{-x})^3$

11. $f(x) = \left(\frac{x^2+3}{x+1}\right)^3$

12. $f(x) = \ln\left(1 + \sqrt{x}\right)$

13. $f(x) = \left(\frac{x^4-x^5}{x^2+x^3}\right)^3$

14. $f(x) = \ln(x^2)$

15. $f(x) = \left(\sqrt{x^3 + 4}\right)^3$

16. $f(x) = \ln\left(e^{x^2}\right)$

17. $f(x) = \left(\frac{x-3}{x+8}\right)^4$

18. $f(x) = e^{x^3}$

19. $f(x) = \left(\frac{x^2-4}{x^3+7}\right)^5$

20. $f(x) = (x^2 + 3)^4$

21. $f(x) = \ln(3x)$

22. $f(x) = \frac{1}{\sqrt[3]{x-1}}$

23. $f(x) = e^{\ln(x^2)}$

24. $f(x) = (x^2 + x)^2(-x^2 + x^3)^{\frac{3}{2}}$

25. $f(x) = \frac{1}{(\sqrt[3]{x}+x^2)^{-2}}$

26. $f(x) = \frac{1}{\sqrt{(x^3+x)^5}}$

27. $f(x) = e^{e^x}$

28. $f(x) = \left(x^{\frac{2}{3}} + x^{\frac{1}{2}}\right)^3$

29. $f(x) = \ln(2x)$

30. $f(x) = (x^3 + x^2 + 2)^5$

31. $f(x) = \ln^4(3x^5)$

32. $f(x) = \sqrt{x^2 + 3}$

33. $f(x) = e^{\ln(x)}$

34. $f(x) = (x^2 + 2)^{-3}$

35. $f(x) = \ln(\ln(x))$

36. $f(x) = (x^2 + 2)^{\frac{3}{2}}$

37. $f(x) = \left(\sqrt{x} + \sqrt[3]{x^2}\right)^{\frac{4}{3}}$

38. $f(x) = e^{x^2}$

39. $f(x) = \sqrt[4]{x^3 + x^2 + 4}$

40. $f(x) = \frac{1}{\sqrt{e^{2x}-e^{3x}}}$

41. $f(x) = (\ln(2x) + \ln(x))^3$

42. $f(x) = \frac{1}{(x^5+x^2)^3}$

43. $f(x) = \ln^3(5x)$

44. $f(x) = \frac{1}{(x^5+x^2)^{\frac{3}{7}}}$

45. $f(x) = (x^3 + 1)^3(5 + x^2)^4$

23

46. $f(x) = \dfrac{1}{(x^4+x)^{\frac{5}{6}}}$

47. $f(x) = \left(\sqrt[3]{x^5} + \sqrt[5]{x^{\frac{9}{2}}}\right)^{\frac{4}{3}}$

48. $f(x) = e^{3x}$

49. $f(x) = \dfrac{1}{\sqrt[5]{\left(\sqrt{x}+x^{\frac{1}{3}}\right)^2}}$

50. $f(x) = \ln(\ln(2x))$

51. $f(x) = \left(x^{\frac{3}{4}} + x^{\frac{1}{2}}\right)^{-4}$

52. $f(x) = \ln(e^x)$

53. $f(x) = (x^2 + 3)^4(x^2 + 2)^{\frac{3}{2}}$

54. $f(x) = (x^3 + x^2 + 1)^{-\frac{1}{2}}$

55. $f(x) = \ln{}^2(x)$

56. $f(x) = \dfrac{1}{\sqrt[4]{\sqrt{x}+x}}$

1. $f'(x) = \frac{2(2x+1)}{3\sqrt[3]{x^2+x}}$

2. $f'(x) = 2e^{2x+e^{2x}}$

3. $f'(x) = \frac{3}{2}x^2(x^3+1)^{-\frac{1}{2}}(x^2+1)^4 + 8x(x^3+1)^{\frac{1}{2}}(x^2+1)^3$

4. $f'(x) = \frac{2}{x}$

5. $f'(x) = \frac{1}{4}\left(x^{\frac{1}{2}} + x^{\frac{4}{3}}\right)^{-\frac{3}{4}}\left(\frac{1}{2}x^{-\frac{1}{2}} + \frac{4}{3}x^{\frac{1}{3}}\right)$

6. $f'(x) = -2\left(\ln^2(x^5) - \ln(x)\right)^{-3}\left(\frac{10\ln(x^5)+1}{x}\right)$

7. $f'(x) = \frac{2}{5}e^{\frac{2}{5}x}$

8. $f'(x) = \frac{2}{15}x^{-\frac{1}{3}}\left(1 + x^{\frac{2}{3}}\right)^{-\frac{4}{5}}$

9. $f'(x) = \frac{5x}{3}(3x+2)\sqrt[3]{(x^3+x^2+4)^2}$

10. $f'(x) = 3(e^x + e^{-x})^2(e^x - e^{-x})$

11. $f'(x) = 3\left(\frac{x^2+3}{x+1}\right)^2\left(\frac{(x+3)(x-1)}{(x+1)^2}\right)$

12. $f'(x) = \frac{1}{2\sqrt{x}(1+\sqrt{x})}$

13. $f'(x) = 3\left(\frac{x^4-x^5}{x^2+x^3}\right)^2\left(\frac{x^5(4-5x)(1+x)-x^5(1-x)(2+3x)}{(x^2+x^3)^2}\right)$

14. $f'(x) = \frac{2}{x}$

15. $f'(x) = \frac{9}{2}x^2\sqrt{x^3+4}$

16. $f'(x) = 2x$

17. $f'(x) = 44\left(\frac{x-3}{x+8}\right)^3$

18. $f'(x) = 3x^2e^{x^3}$

19. $f'(x) = 5\left(\frac{x^2-4}{x^3+7}\right)^4\left(\frac{-x^4+14x+12x^2}{(x^3+7)^2}\right)$

20. $f'(x) = 8x(x^2+3)^3$

21. $f'(x) = \frac{1}{x}$

22. $f'(x) = \frac{-1}{3\sqrt[3]{(x-1)^4}}$

23. $f'(x) = 2x$

24. $f'(x) = 2(x^2+x)^1(2x+1)(-x^2+x^3)^{\frac{3}{2}} + (x^2+x)^2\frac{3}{2}(-x^2+x^3)^{\frac{1}{2}}(-2x+3x)^2$

25. $f'(x) = 2\left(x^{\frac{1}{3}} + x^2\right)\left(\frac{1}{3}x^{-\frac{2}{3}} + 2x\right)$

26. $f'(x) = \frac{-5(3x^2+1)}{2\sqrt{(x^3+x)^7}}$

27. $f'(x) = e^{x+e^x}$

28. $f'(x) = 3\left(x^{\frac{2}{3}} + x^{\frac{1}{2}}\right)^2\left(\frac{2}{3}x^{-\frac{1}{3}} + \frac{1}{2}x^{-\frac{1}{2}}\right)$

29. $f'(x) = \frac{1}{x}$

30. $f'(x) = 5x(x^3+x^2+2)^4(3x+2)$

31. $f'(x) = \frac{20\ln^3(3x^5)}{x}$

32. $f'(x) = \frac{x}{\sqrt{x^2+3}}$

33. $f'(x) = 1$

34. $f'(x) = \dfrac{-6x}{(x^2+2)^4}$

35. $f'(x) = \dfrac{1}{x\ln(x)}$

36. $f'(x) = 3x\sqrt{x^2+2}$

37. $f'(x) = \dfrac{4}{3}\left(x^{\frac{1}{2}} + x^{\frac{2}{3}}\right)^{\frac{1}{3}}\left(\dfrac{1}{2}x^{-\frac{1}{2}} + \dfrac{2}{3}x^{-\frac{1}{3}}\right)$

38. $f'(x) = 2xe^{x^2}$

39. $f'(x) = \dfrac{x(3x+2)}{4\sqrt[4]{(x^3+x^2+4)^3}}$

40. $f'(x) = -\dfrac{1}{2}e^{2x}(2 - 3e^x)(e^{2x} - e^{3x})^{-\frac{3}{2}}$

41. $f'(x) = \dfrac{6(\ln(2x)+\ln(x))^2}{x}$

42. $f'(x) = \dfrac{-3x(5x^3+2)}{(x^5+x^2)^4}$

43. $f'(x) = \dfrac{3\ln^2(5x)}{x}$

44. $f'(x) = \dfrac{-3x(5x^3+2)}{7\sqrt[7]{(x^5+x^2)^{10}}}$

45. $f'(x) = 9x^2(x^3+1)^2(5+x^2)^4 + 8x(x^3+1)^3(5+x^2)^3$

46. $f'(x) = \dfrac{-5(4x^3+1)}{6\sqrt[6]{(x^4+x)^{11}}}$

47. $f'(x) = \dfrac{4}{3}\left(x^{\frac{5}{3}} + x^{\frac{9}{10}}\right)^{\frac{1}{3}}\left(\dfrac{5}{3}x^{\frac{2}{3}} + \dfrac{9}{10}x^{-\frac{1}{10}}\right)$

48. $f'(x) = 3e^{3x}$

49. $f'(x) = -\dfrac{2}{5}\left(x^{\frac{1}{2}} + x^{\frac{1}{3}}\right)^{-\frac{7}{5}}\left(\dfrac{1}{2}x^{-\frac{1}{2}} + \dfrac{1}{3}x^{-\frac{2}{3}}\right)$

50. $f'(x) = \dfrac{1}{x\ln(x)}$

51. $f'(x) = -4\left(x^{\frac{3}{4}} + x^{\frac{1}{2}}\right)^{-5}\left(\dfrac{3}{4}x^{-\frac{1}{4}} + \dfrac{1}{2}x^{-\frac{1}{2}}\right)$

52. $f'(x) = 1$

53. $f'(x) = 8x(x^2+3)^3(x^2+2)^{\frac{3}{2}} + 3x(x^2+3)^4(x^2+2)^{\frac{1}{2}}$

54. $f'(x) = \dfrac{-x(3x+2)}{2\sqrt{(x^3+x^2+1)^3}}$

55. $f'(x) = \dfrac{2\ln(x)}{x}$

56. $f'(x) = -\dfrac{1}{4}\left(x^{\frac{1}{2}} + x\right)^{-\frac{5}{4}}\left(\dfrac{1}{2}x^{-\frac{1}{2}} + 1\right)$

Find the derivative $f'(x)$ of each of the following functions.

1. $f(x) = \sin(e^x)$

2. $f(x) = e^{e^{2x}}$

3. $f(x) = \sqrt[3]{(x^2 + x)^2}$

4. $f(x) = e^{\arccos(2x)}$

5. $f(x) = \ln(3x^2)$

6. $f(x) = \sin^2(x)$

7. $f(x) = (x^3 + x^2 + 4)^{\frac{5}{3}}$

8. $f(x) = \ln(\arctan(x))$

9. $f(x) = \sqrt{x^3 + 1}(x^2 + 1)^4$

10. $f(x) = \arctan\left(\sin\left(\ln\left(e^{\sqrt{x}}\right)\right)\right)$

11. $f(x) = \frac{1}{\sqrt[3]{x-1}}$

12. $f(x) = \frac{e^{2x}}{\sin(3x)}$

13. $f(x) = \sqrt[5]{1 + x^{\frac{2}{3}}}$

14. $f(x) = \cos\left(x^{\frac{3}{4}}\right)$

15. $f(x) = \left(\ln^2(x^5) - \ln(x)\right)^{-2}$

16. $f(x) = \arcsin(\arccos(x))$

17. $f(x) = \left(\frac{x^2 - 4}{x^3 + 7}\right)^5$

18. $f(x) = \sin^3(\cos(2x))$

19. $f(x) = \ln(x^2)$

20. $f(x) = \arctan(3x)$

21. $f(x) = e^{2x}\tan^3(x)$

22. $f(x) = e^{\frac{2}{5}x}$

23. $f(x) = \arcsin^3(x^2)$

24. $f(x) = \ln(3x)$

25. $f(x) = \sin\left(\sin^2\left(\sin^3(x^4)\right)\right)$

26. $f(x) = (e^x + e^{-x})^3$

27. $f(x) = \tan^4(x)$

28. $f(x) = \ln(1 + \sqrt{x})$

29. $f(x) = \ln\left(e^{x^2}\right)$

30. $f(x) = \sin(\arctan(x))$

31. $f(x) = \left(\frac{x^4 - x^5}{x^2 + x^3}\right)^3$

32. $f(x) = \ln^4\left(\cos\left(e^{\sin(x^2)}\right)\right)$

33. $f(x) = \left(\frac{x^2 + 3}{x + 1}\right)^3$

34. $f(x) = \sin(3x)$

35. $f(x) = \sqrt[4]{\sqrt{x} + x^{\frac{4}{3}}}$

36. $f(x) = \ln(\sin(x))$

37. $f(x) = e^{x^3}$

38. $f(x) = \cos^5(x^2)$

39. $f(x) = e^{e^x}$

40. $f(x) = \tan(3x)$

41. $f(x) = \frac{1}{\left(\sqrt[3]{x} + x^2\right)^{-2}}$

42. $f(x) = \arctan(\sin(x))$

43. $f(x) = \left(\sqrt{x^3 + 4}\right)^3$

44. $f(x) = e^{\cos(4x)}$

45. $f(x) = (x^2 + x)^2(-x^2 + x^3)^{\frac{3}{2}}$

46. $f(x) = \frac{1}{\sqrt{(x^3 + x)^5}}$

47. $f(x) = \left(\frac{x-3}{x+8}\right)^4$

48. $f(x) = \tan(\sin(x))$

49. $f(x) = e^{\ln(x^2)}$

50. $f(x) = \arctan{}^2(3x)$

51. $f(x) = (x^2 + 3)^4$

52. $f(x) = \dfrac{\ln(4x)}{\arctan(3x)}$

53. $f(x) = (\ln(2x) + \ln(x))^3$

54. $f(x) = \left(\sqrt{x} + \sqrt[3]{x^2}\right)^{\frac{4}{3}}$

55. $f(x) = \sin{}^5(x^2)$

56. $f(x) = \ln(2x)$

57. $f(x) = \arctan(\arcsin(x))$

58. $f(x) = \dfrac{1}{\sqrt[4]{\sqrt{x}+x}}$

59. $f(x) = e^{3x}$

60. $f(x) = e^{\tan(\ln(3x))}$

61. $f(x) = \sqrt{x^2 + 3}$

62. $f(x) = \cos{}^3(x)$

63. $f(x) = \left(x^{\frac{3}{4}} + x^{\frac{1}{2}}\right)^{-4}$

64. $f(x) = \tan(\arcsin(x))$

65. $f(x) = \ln(\ln(x))$

66. $f(x) = \cos\left(\dfrac{1}{2}x\right)$

67. $f(x) = (x^2 + 2)^{\frac{3}{2}}$

68. $f(x) = \dfrac{1}{(x^5+x^2)^{\frac{3}{7}}}$

69. $f(x) = e^{\arctan(x)}$

70. $f(x) = \ln(e^x)$

71. $f(x) = \arctan{}^4(\cos(\ln(5x)))$

72. $f(x) = (x^3 + x^2 + 2)^5$

73. $f(x) = \ln(\arcsin(x))$

74. $f(x) = \tan(e^{3x})$

75. $f(x) = \left(x^{\frac{2}{3}} + x^{\frac{1}{2}}\right)^3$

76. $f(x) = e^{x^2}$

77. $f(x) = \sin(2x)\cos(3x)$

78. $f(x) = \dfrac{1}{(x^4+x)^{\frac{5}{6}}}$

79. $f(x) = \ln{}^4(3x^5)$

80. $f(x) = \tan(x^5)$

81. $f(x) = \dfrac{1}{\sqrt[5]{\left(\sqrt{x}+x^{\frac{1}{3}}\right)^2}}$

82. $f(x) = \ln{}^3(5x)$

83. $f(x) = \tan{}^4\left(\ln\left(e^{\sin(3x)}\right)\right)$

84. $f(x) = (x^2 + 2)^{-3}$

85. $f(x) = (x^3 + x^2 + 1)^{-\frac{1}{2}}$

86. $f(x) = e^{\ln(x)}$

87. $f(x) = \arcsin(\tan(x))$

88. $f(x) = \ln(\ln(2x))$

89. $f(x) = \arcsin^3(x)$

90. $f(x) = \ln^2(x)$

91. $f(x) = (x^3 + 1)^3(5 + x^2)^4$

92. $f(x) = \frac{1}{(x^5 + x^2)^3}$

93. $f(x) = \sin(x^2)$

94. $f(x) = \tan^4(x^5)$

95. $f(x) = \frac{1}{\sqrt{e^{2x} - e^{3x}}}$

96. $f(x) = \arccos(e^x)$

97. $f(x) = \sqrt[4]{x^3 + x^2 + 4}$

98. $f(x) = (x^2 + 3)^4(x^2 + 2)^{\frac{3}{2}}$

99. $f(x) = \left(\sqrt[3]{x^5} + \sqrt[5]{x^{\frac{9}{2}}}\right)^{\frac{4}{3}}$

100. $f(x) = \arctan^2(x)$

101. $f(x) = e^{\ln(\sin(x))}$

1. $f'(x) = e^x \cos(e^x)$

2. $f'(x) = 2e^{2x+e^{2x}}$

3. $f'(x) = \frac{2(2x+1)}{3\sqrt[3]{x^2+x}}$

4. $f'(x) = \frac{2e^{\arcsin(2x)}}{\sqrt{1-4x^2}}$

5. $f'(x) = \frac{2}{x}$

6. $f'(x) = 2\sin(x)\cos(x)$

7. $f'(x) = \frac{5x}{3}(3x+2)\sqrt[3]{(x^3+x^2+4)^2}$

8. $f'(x) = \frac{1}{(1+x^2)\arctan(x)}$

9. $f'(x) = \frac{3}{2}x^2(x^3+1)^{-\frac{1}{2}}(x^2+1)^4 + 8x(x^3+1)^{\frac{1}{2}}(x^2+1)^3$

10. $f'(x) = \frac{\cos(\ln(e^{\sqrt{x}}))}{2\sqrt{x}(1+\sin^2(\ln(e^{\sqrt{x}})))}$

11. $f'(x) = \frac{-1}{3\sqrt[3]{(x-1)^4}}$

12. $f'(x) = \frac{2e^{2x}\sin(3x)+3\cos(3x)e^{2x}}{\sin^2(3x)}$

13. $f'(x) = \frac{2}{15}x^{-\frac{1}{3}}\left(1+x^{\frac{2}{3}}\right)^{-\frac{4}{5}}$

14. $f'(x) = -\frac{3}{4}x^{\frac{-1}{4}}\sin\left(x^{\frac{3}{4}}\right)$

15. $f'(x) = -2\left(\ln^2(x^5)-\ln(x)\right)^{-3}\left(\frac{10\ln(x^5)+1}{x}\right)$

16. $f'(x) = \frac{1}{\sqrt{1-\arccos^2(x)}}\left(\frac{-1}{\sqrt{1-x^2}}\right)$

17. $f'(x) = 5\left(\frac{x^2-4}{x^3+7}\right)^4\left(\frac{-x^4+14x+12x^2}{(x^3+7)^2}\right)$

18. $f'(x) = -6\sin(2x)\cos(\cos(2x))\sin^2(\cos(2x))$

19. $f'(x) = \frac{2}{x}$

20. $f'(x) = \frac{3}{1+9x^2}$

21. $f'(x) = e^{2x}\tan^2(x)\left[2\tan(x)+3\sec^2(x)\right]$

22. $f'(x) = \frac{2}{5}e^{\frac{2}{5}x}$

23. $f'(x) = \frac{6x\arcsin^2(x^2)}{\sqrt{1-x^4}}$

24. $f'(x) = \frac{1}{x}$

25. $f'(x) =$
$24x^3\cos\left(\sin^2(\sin^3(x^4))\right)\left(\sin(\sin^3(x^4))\right)\left(\cos(\sin^3(x^4))\right)\left(\sin^2(x^4)\right)\left(\cos(x^4)\right)$

26. $f'(x) = 3(e^x+e^{-x})^2(e^x-e^{-x})$

27. $f'(x) = 4\tan^3(x)\sec^2(x)$

28. $f'(x) = \frac{1}{2\sqrt{x}(1+\sqrt{x})}$

29. $f'(x) = 2x$

30. $f'(x) = \frac{\cos(\arctan(x))}{1+x^2}$

31. $f'(x) = 3\left(\frac{x^4-x^5}{x^2+x^3}\right)^2\left(\frac{x^5(4-5x)(1+x)-x^5(1-x)(2+3x)}{(x^2+x^3)^2}\right)$

32. $f'(x) = -8xe^{\sin(x^2)}\cos(x^2)\tan\left(e^{\sin(x^2)}\right)\ln^3\left(\cos\left(e^{\sin(x^2)}\right)\right)$

33. $f'(x) = 3\left(\frac{x^2+3}{x+1}\right)^2 \left(\frac{(x+3)(x-1)}{(x+1)^2}\right)$

34. $f'(x) = 3\cos(3x)$

35. $f'(x) = \frac{1}{4}\left(x^{\frac{1}{2}} + x^{\frac{4}{3}}\right)^{-\frac{3}{4}}\left(\frac{1}{2}x^{-\frac{1}{2}} + \frac{4}{3}x^{\frac{1}{3}}\right)$

36. $f'(x) = \cot(x)$

37. $f'(x) = 3x^2 e^{x^3}$

38. $f'(x) = -10x\cos^4(x^2)\sin(x^2)$

39. $f'(x) = e^{x+e^x}$

40. $f'(x) = 3\sec^2(3x)$

41. $f'(x) = 2\left(x^{\frac{1}{3}} + x^2\right)\left(\frac{1}{3}x^{-\frac{2}{3}} + 2x\right)$

42. $f'(x) = \frac{\cos(x)}{1+\sin^2(x)}$

43. $f'(x) = \frac{9}{2}x^2\sqrt{x^3+4}$

44. $f'(x) = -4\sin(4x)e^{\cos(4x)}$

45. $f'(x) = 2(x^2+x)^1(2x+1)(-x^2+x^3)^{\frac{3}{2}} +$
$\qquad +(x^2+x)^2\frac{3}{2}(-x^2+x^3)^{\frac{1}{2}}(-2x+3x)^2$

46. $f'(x) = \frac{-5(3x^2+1)}{2\sqrt{(x^3+x)^7}}$

47. $f'(x) = 44\left(\frac{x-3}{x+8}\right)^3$

48. $f'(x) = \sec^2(\sin(x))\cos(x)$

49. $f'(x) = 2x$

50. $f'(x) = \frac{6\arctan(x)}{1+9x^2}$

51. $f'(x) = 8x(x^2+3)^3$

52. $f'(x) = \frac{\frac{\arctan(3x)}{x} - \frac{3\ln(4x)}{1+9x^2}}{\arctan^2(3x)}$

53. $f'(x) = \frac{6(\ln(2x)+\ln(x))^2}{x}$

54. $f'(x) = \frac{4}{3}\left(x^{\frac{1}{2}} + x^{\frac{2}{3}}\right)^{\frac{1}{3}}\left(\frac{1}{2}x^{-\frac{1}{2}} + \frac{2}{3}x^{-\frac{1}{3}}\right)$

55. $f'(x) = 10x\sin^4(x^2)\cos(x^2)$

56. $f'(x) = \frac{1}{x}$

57. $f'(x) = \frac{1}{1+\arcsin^2(x)}\left(\frac{1}{\sqrt{1-x^2}}\right)$

58. $f'(x) = -\frac{1}{4}\left(x^{\frac{1}{2}} + x\right)^{-\frac{5}{4}}\left(\frac{1}{2}x^{-\frac{1}{2}} + 1\right)$

59. $f'(x) = 3e^{3x}$

60. $f'(x) = \frac{e^{\tan(\ln(3x))}\sec^2(\ln(3x))}{x}$

61. $f'(x) = \frac{x}{\sqrt{x^2+3}}$

62. $f'(x) = -3\cos^2(x)\sin(x)$

63. $f'(x) = -4\left(x^{\frac{3}{4}} + x^{\frac{1}{2}}\right)^{-5}\left(\frac{3}{4}x^{-\frac{1}{4}} + \frac{1}{2}x^{-\frac{1}{2}}\right)$

64. $f'(x) = \frac{\sec^2(\arcsin(x))}{\sqrt{1-x^2}}$

65. $f'(x) = \frac{1}{x\ln(x)}$

66. $f'(x) = -\frac{1}{2}\sin\left(\frac{1}{2}x\right)$

67. $f'(x) = 3x\sqrt{x^2 + 2}$

68. $f'(x) = \dfrac{-3x(5x^3+2)}{7\sqrt[7]{(x^5+x^2)^{10}}}$

69. $f'(x) = \dfrac{e^{\arctan(x)}}{1+x^2}$

70. $f'(x) = 1$

71. $f'(x) = \dfrac{-20\sin(\ln(5x))\arctan^3(\cos(\ln(5x)))}{x(1+\cos^2(\ln(5x)))}$

72. $f'(x) = 5x(x^3 + x^2 + 2)^4(3x + 2)$

73. $f'(x) = \dfrac{1}{\sqrt{1-x^2}\arcsin(x)}$

74. $f'(x) = e^{3x}\sec^2(e^{3x})$

75. $f'(x) = 3\left(x^{\frac{2}{3}} + x^{\frac{1}{2}}\right)^2 \left(\frac{2}{3}x^{-\frac{1}{3}} + \frac{1}{2}x^{-\frac{1}{2}}\right)$

76. $f'(x) = 2xe^{x^2}$

77. $f'(x) = 2\cos(2x)\cos(3x) - 3\sin(2x)\sin(3x).$

78. $f'(x) = \dfrac{-5(4x^3+1)}{6\sqrt[6]{(x^4+x)^{11}}}$

79. $f'(x) = \dfrac{20\ln^3(3x^5)}{x}$

80. $f'(x) = 5x^4\sec^2(x^5)$

81. $f'(x) = -\frac{2}{5}\left(x^{\frac{1}{2}} + x^{\frac{1}{3}}\right)^{-\frac{7}{5}} \left(\frac{1}{2}x^{-\frac{1}{2}} + \frac{1}{3}x^{-\frac{2}{3}}\right)$

82. $f'(x) = \dfrac{3\ln^2(5x)}{x}$

83. $f'(x) = 12\cos(3x)\tan^3\left(\ln\left(e^{\sin(3x)}\right)\right)\sec^2\left(\ln\left(e^{\sin(3x)}\right)\right)$

84. $f'(x) = \dfrac{-6x}{(x^2+2)^4}$

85. $f'(x) = \dfrac{-x(3x+2)}{2\sqrt{(x^3+x^2+1)^3}}$

86. $f'(x) = 1$

87. $f'(x) = \dfrac{\sec^2(x)}{\sqrt{1-\tan^2(x)}}$

88. $f'(x) = \dfrac{1}{x\ln(x)}$

89. $f'(x) = \dfrac{3\arcsin^2(x)}{\sqrt{1-x^2}}$

90. $f'(x) = \dfrac{2\ln(x)}{x}$

91. $f'(x) = 9x^2(x^3 + 1)^2(5 + x^2)^4 + 8x(x^3 + 1)^3(5 + x^2)^3$

92. $f'(x) = \dfrac{-3x(5x^3+2)}{(x^5+x^2)^4}$

93. $f'(x) = 2x\cos(x^2)$

94. $f'(x) = 20x^4\tan^3(x^5)\sec^2(x^5)$

95. $f'(x) = -\frac{1}{2}e^{2x}(2-3e^x)(e^{2x}-e^{3x})^{-\frac{3}{2}}$

96. $f'(x) = \frac{-e^x}{\sqrt{1-e^{2x}}}$

97. $f'(x) = \frac{x(3x+2)}{4\sqrt[4]{(x^3+x^2+4)^3}}$

98. $f'(x) = 8x(x^2+3)^3(x^2+2)^{\frac{3}{2}} + 3x(x^2+3)^4(x^2+2)^{\frac{1}{2}}$

99. $f'(x) = \frac{4}{3}\left(x^{\frac{5}{3}} + x^{\frac{9}{10}}\right)^{\frac{1}{3}}\left(\frac{5}{3}x^{\frac{2}{3}} + \frac{9}{10}x^{-\frac{1}{10}}\right)$

100. $f'(x) = \frac{2\arctan(x)}{1+x^2}$

101. $f'(x) = \cos(x)$

"Only he who never plays, never loses"